wellbeing
at v.

how to manage workplace
wellness to boost your staff
and business performance

consultant editor:	Marc Beishon
sub-editor:	Lesley Malachowski
production manager:	Lisa Robertson
design:	Halo Design
head of commercial relations:	Nicola Morris
commercial director:	Sarah Ready
publishing director:	Tom Nash
chief operating officer:	Andrew Main Wilson

Published for the Institute of Directors, Standard Life
Healthcare, the Department for Work and Pensions (DWP)
and the Health and Safety Executive (HSE) by Director
Publications Ltd, 116 Pall Mall London SW1Y 5ED
Ⓣ 020 7766 8950 Ⓦ www.iod.com

·blications Ltd
om the British Library

Standard Life Healthcare

Standard Life Healthcare is an award-winning private medical insurer, and one of the largest providers in the UK. We supply a range of healthcare plans to corporate and individual customers. As well as helping people to access the treatment they need at a time convenient to them, our products also support the health and wellbeing of our customers – we focus on wellness, not just illness. We provide free access to an online health and wellbeing service and a free 24 hour GP advice line. We have a reputation for excellent customer service, winning the Health Insurance Best Customer Service award for five successive years.

Health and Safety Executive

The Health and Safety Executive (HSE) is the government agency responsible for health and safety at work in Britain. It provides inspection, advice and information on how to manage health and safety within organisations and guidance to employers on how to meet their legal responsibilities and duties to workers and the public. HSE has teams of inspectors and technical experts in 20 offices across the country and runs advertising and information campaigns on major health and safety issues such as stress and back pain, as well as providing printed material and an extensive website featuring information and guidance for business, workers and the public.

Department for Work and Pensions

The Department for Work and Pensions (DWP) is here to:

- ☐ promote opportunity and independence for all
- ☐ help individuals achieve their potential through employment
- ☐ work to end poverty in all its forms

CONTENTS

Private medical insurance can benefit your staff *and* your company.

STANDARD LIFE

Whether you run a consultancy employing a select few or a national sales force all over the country, people will be at the heart of your organisation's ability to compete and succeed.

A Standard Life private medical insurance plan will ensure your staff get treated as soon as possible should they fall ill, reducing the cost of absenteeism to your business. They'll see the advantage of being able to take control of their health – and you'll see it rewarded in a happy and loyal workforce – and improved productivity.

Standard Life deliver healthcare solutions that suit todays' businesses – bringing you outstanding cover at our best possible price.

To find out more call
0800 333 312

For the financial realities of life

Pensions   Mortgages   Savings   Investments   **Healthcare**   Insurance

striving for healthy performance

**Miles Templeman, Director General
Institute of Directors**

All enterprises seek to be in healthy state. If their employees are in a good state of health and wellbeing that must surely contribute to successful performance. The downside of that is when time and effort is lost through sickness and other absence from work, or when employees are not getting or giving of their best.

When we published a previous Director's Guide on this subject in 2002 my predecessor highlighted the "sea change in thinking about the health and wellbeing of employees". That thinking has continued and the present publication highlights several key topics around health and wellbeing at work.

The IoD has been engaging positively with several of the public policy developments on health and wellbeing at work. These include the government's National Stakeholder Council on Health, Work and Wellbeing, and dialogue with the Department for Work and Pensions and the Health and Safety Executive.

Members of the IoD know all too well that regulations have to be followed and coped with. Yet health and wellbeing are aspects of performance that benefit from a voluntary – and willing – approach rather than a focus solely on rules and regulations. The guide gives an overview of the business case for involvement, together with suggestions for practical action to help employers.

The challenge set out in this guide is to plan effectively and take action around health and wellbeing, thereby creating higher performing workplaces. If that can be achieved it will be good for the individual employee, good for the enterprise and good for the country as a whole.

Does health & safety pay? Octel found just the right formula.

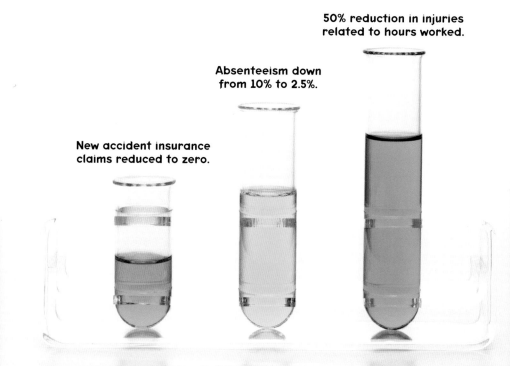

50% reduction in injuries related to hours worked.

Absenteeism down from 10% to 2.5%.

New accident insurance claims reduced to zero.

Does health & safety pay? It did for Associated Octel, manufacturers of petroleum additives and speciality chemicals.

Employees have benefited from fewer accidents and injuries, whilst Octel as a company has enjoyed reduced production costs and a huge fall in insurance claims.

HSE

Better health & safety benefits everyone

To find out how your business could make health & safety pay, visit www.hse.gov.uk or call 0845 345 0055.*

*Lines are open 8am - 6pm Monday to Friday.

healthy dividends

Professor Dame Carol Black
National Director for Health and Work

A business's most valuable asset is, and will always be, the dedicated staff that devote themselves to delivering the work of the organisation. Healthy and fit staff are essential to ensuring a company remains efficient and profitable. Every chief executive and senior manager should be aware of the benefits that can be achieved by simple measures described within this guide.

None of us doubt that good staff management practices ensure that our workforce delivers our aims – but many of us forget that unless we help them manage their health, fitness, and wellbeing many of our workers can and will fall ill. Surveys of our workers show that they value these aspects of their work more than just financial rewards. People want to perform to the best of their ability.

We know that work is good for people. It provides economic stability as well as being a valuable source of social interaction both for the individual and the community within which they work. Fit, healthy staff deliver profitable businesses which in turn allow the UK to remain one of the most prosperous and best places to work and live.

We now have an employment rate of 75 per cent. To remain world class we must ensure that our workers can deliver business's needs. When workers fall ill we must ensure we support them back to health and back to work. We also need to value the skills our older workers have, recognising the strengths and experience they can bring to our businesses and accepting that they may not have the physical capabilities of someone just starting their working career.

This guide provides you with the knowledge and tools to take advantage of the fabulous opportunity to improve the health and wellbeing of your workers – to both their benefit and yours.

IS YOUR ORGANISATION READY FOR AGE LEGISLATION?

FREE BE READY ORGANISER

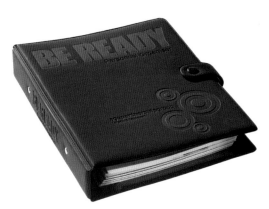

Age legislation is coming in October 2006. Are you ready? Employers of all sectors and sizes are already changing the way they work and reaping the benefits of a mixed-age workforce.

Order your free Organiser today. Content includes:

- Department of Trade & Industry and Acas guidance
- Top tips to help you prepare
- The experts: who to contact and how

Call 0845 715 2000

or email apg@trgeuropeplc.com for your free copy now.

Website: www.agepositive.gov.uk/agepartnershipgroup

Please note: If you have previously ordered a Be Ready Organiser, you will automatically receive updated inserts for your Organiser in the near future.

age**partnership**group
Targeting Employers®

Quote ref: APG 1

setting the scene

With the health of the UK workforce in the spotlight as never before, business and government must work together to address the health and wellbeing of the working population, says Marc Beishon, freelance healthcare writer

Workplace health and wellbeing is rising steadily up the agenda of British business as companies recognise the contribution made by a workforce that is fit and healthy. The evidence is building that concerted workplace health and 'wellness' policies can bring about:

- lower absenteeism and costs associated with ill health

- increased productivity and improved relationships with customers and suppliers

- better staff morale

- higher reputation as an employer and reduced staff turnover

EXECUTIVE SUMMARY

- 28 million working days are lost through ill health each year

- there is plenty of help and support available to companies wanting to introduce practices and policies for wellbeing at work

- better profitability and productivity are among the benefits of providing employees with wellbeing support

Such policies will also contribute to the government's agenda for improving the health of individuals and the economic wellbeing of the country. As a 'setting', the workplace has become a crucial meeting place for a range of issues that concern the government and a wide range of agencies, including trade unions, health and safety agencies, and voluntary organisations such as the British Heart Foundation. The 'headline' issues include:

- promoting workplace health in the light of trends such as rising obesity rates and alcohol consumption, and addressing the needs of older workers and people with special needs who are entering or re-entering employment

- helping people to return to work after long-term absence – an especially critical part of the government's strategy, given the cost to the economy of incapacity benefit

- improving access to occupational health services, especially for smaller firms, to help avoid small health problems turning into long-term absence

The cost and scale of the workplace health problem is certainly not underplayed by the government. Ministers at a recent Health, Work and Wellbeing summit hammered home the core statistics: 35 million working days lost through ill health and injury at a cost of £12bn, and 2.7 million people of working age on benefit. In its latest annual absence survey, the Confederation of British Industry (CBI) puts the total number of days lost through all absence at the lowest since its first survey in 1987. But one shouldn't be deceived. In its estimation, absence costs have risen to £13bn, and a 'yawning gap' of almost nine days a year exists between the best and worst performing organisations.

Drilling deeper, the Work Foundation asserts that "there is no doubt that the health of the UK workforce is getting worse". It claims that:

- over 25 per cent of UK workers have a long-standing condition that affects their ability to work

- over half of all days lost through sickness absence are accounted for by long-term absences

- the proportion of the workforce with common mental health problems (depression, anxiety, etc.) is now 16 per cent

- while between three and five per cent of the UK workforce may be away from work owing to sickness on any one day (depending on the industry or sector), a further 25 per cent to 30 per cent are at work but performing sub-optimally because of ill health

Other commentators argue over such statistics, but invariably accept that employers face challenges in addressing workplace health, especially smaller firms without an occupational health department or access to such expertise. Larger organisations have pioneered many aspects of health and wellbeing management. AstraZeneca, for example, employs a wellbeing manager, while BT is known for its flexible working practices and, of course, the NHS has the huge Improving Working Lives programme. (Some public sector organisations have been taking a lead, thanks to a more receptive culture – and an oft-cited pressing need to address high absence rates. However, it is worth noting that the HSE recently emphasised that under-reporting of absence in the private sector – particularly by SMEs – coupled with differences in workforce demographics, means that the public/private difference is actually quite small).

Across the board – and especially in smaller firms – there are obstacles and differing priorities. A recent survey by Standard Life Healthcare found that safety remains the most important priority – preventing accidents and avoiding litigation. But, 82 per cent of firms recognise that good health and wellbeing support provides benefits such as better profitability and productivity.

The survey reveals, however, a possible blind spot in this thinking. The items on the top of companies' 'wish lists' for providing health services are making private medical insurance tax deductible, grant programmes for health and wellbeing initiatives, and other company tax breaks. This suggests that employers are not yet convinced that 'wellness' programmes are cost-effective in their own right.

But 57 per cent of those surveyed are working in some way to cut the costs of ill health (and a further 12.5 per cent would like to do so, but don't know how). According to the Standard Life survey, the support currently provided is:

- absence management (long term)   64 per cent
- absence management (short term)   63 per cent
- private medical insurance   52 per cent
- employee assistance programmes   37 per cent
- preventive health (eg. free fruit, healthy restaurant)   22 per cent
- cash plans   10 per cent
- none of these   11 per cent

About half spend between £0-100 a year on health per employee. However, most stated that if they had more to spend, they would opt to put more into preventive measures such as healthy eating and gym membership. Indeed, research conducted by Business in the Community in 2005 – the Spend Now, Save

Now survey – found that many CEOs and financial directors believe it is more important to improve employees' energy and alertness than to offer performance-related pay. There is growing awareness, too, of the underlying structural reasons for poor wellbeing at work, such as job design, the use of correct absence management procedures and health profiling tools.

There has never been a stronger focus on workplace health from various quarters, including the government. There are now many 'enabling' programmes and regulations, some of which can help make the kind of practical steps that companies will need to help fill the holes in occupational health provision in Britain. Much of this emphasis was signaled in *Choosing Health*, the public health White Paper of 2004, which devoted a section to workplace health. As the document notes:

"Workplaces are often underutilised as a setting for promoting health and wellbeing. This is something that individual employees cannot achieve for themselves but need assistance from employers, government and trade unions."

Among the array of announcements and regulation in the last few years have been:

- the publication in 2005 of *Health, work and wellbeing – caring for our future*, a commendably brief document that sets out a government strategy on the key themes of engaging stakeholders, improving working lives and providing healthcare for working age people

- the appointment of the first ever national director for Health and Work – currently Dame Carol Black – and the creation of a national stakeholder council and charter for the Health, work and wellbeing strategy

- the establishment of Workplace Health Connect in 2006 by the HSE, a health at work service for smaller companies (see also chapter 10)

- the Pathways to Work programme and the Welfare Reform Bill, both of which aim to help people back into employment

- the publication by the HSE of Stress Management Standards, and other HSE projects such as a worker risk involvement programme and campaigns such as Backs! (for back pain)

- the ban on workplace smoking in England, Wales and Northern Ireland, in 2007 (the ban is already in place in Scotland)

- regulations on flexible working for parents and carers, and the European Working Time Directive

- the Disability Discrimination Act, which addressees the working rights of disabled people

There are several other initiatives worth noting that are outside, or allied with, government. These include:

- a Health and Wellbeing at Work project for the Investors in People Standard. This is currently being piloted (see also chapter 2)

- Well@Work – a joint initiative by the British Heart Foundation, Sport England, the Big Lottery Fund and the Department of Health, to test ways of making workplaces healthier and more 'active'

- the creation of workplace health and wellness assessment tools by organisations such as Standard Life Healthcare, Wellkom and the Work Foundation, and also the HSE (see also chapters 2 and 3)

- Business in the Community's Action on Health programme, which includes the aim of establishing reporting on health as commonplace in UK boardrooms, and which also has a healthy workplace award scheme

When you start looking for information on workplace health, it is obvious that there are many such initiatives, to the extent that it can get a little confusing. It is the purpose of this guide to break out the key elements into digestible sections and also to provide a list that will help ensure important resources are not missed (see also resource section at the back of the guide).

The claim that good health and safety practice can contribute to the bottom-line does not need to be taken on faith. There is an increasing body of evidence that not only includes raw global figures on absenteeism rates, but also highlights the links between health and productivity, plus detailed case studies on how even smaller companies are benefiting (see also chapters 2 and 4).

It may also be salutary to know that some institutional investors are paying attention to the health and safety performance of firms they are interested in.

A report in 2002 for the HSE, for example, suggested that six indicators could be used to judge the performance of a company:

- whether a director has been named as a health and safety champion
- the level of reporting of health and safety management systems
- the number of fatalities
- the time lost to injury
- the absenteeism rate
- the cost of health and safety losses

Not all of these indicators are universally accepted. The IoD, for example, argues that the first point is not necessary to demonstrate good management – and tends to mix up the role of direction and management. Nevertheless – used formally or informally – such indicators can only add to the picture of good management, and will be increasingly seen as an important part of the overall corporate social responsibility agenda and the rise of interest in socially responsible investment.

what does a healthy company look like?

> You don't need access to expensive schemes or resources in order to be a healthy organisation, says Marc Beishon. It's more to do with addressing the causes of discontent

The question in the title of this chapter is intended to provoke. No doubt, every person who is asked it will have a different answer, from healthy canteens and an in-house gym to a low absence rate, to a high percentage of flexible workers and a workers' council. In fact, all such ideas can have a valid place to a greater or lesser extent in a company that has chosen to make health and wellbeing a central plank of its operation and culture.

As a growing body of work indicates, what is important is to have a framework that draws together the elements that are most appropriate to the company. And, while the ideal is that all elements are represented, some are more important than others, particularly in the initial stages of introducing and then reinforcing good practice.

EXECUTIVE SUMMARY

- ☐ Investors in People has developed a framework that challenges firms to become more aware of health and wellbeing issues

- ☐ according to one survey, the nature of the work, work environment and working relationships are far more important than pay in determining job satisfaction

This is certainly the thinking behind the new Health and Wellbeing at Work project being piloted by Investors in People, and authorities such as the Work Foundation. A point that comes through loud and clear is that there is a big difference between 'good work' and 'bad work'. Poor quality employment, says the Work Foundation, "is associated with low levels of wellbeing, a higher incidence of physical or mental illness, low levels of self-esteem and a sense of powerlessness". Job design, involvement in decision-making, managerial competence, bullying and harassment, and status are among the factors determining the quality of the work experience. The links between poor health and lower status and job control have been famously explored in Professor Michael Marmot's Whitehall studies of civil servants.

As a resource pack for the Health and Wellbeing at Work project notes, many employers tend to think first of support offerings such as gym membership and medical insurance when it comes to workplace health. Yet, it states: "Some of the 'healthiest' organisations are smaller ones without the resources or flexibility to support expensive schemes, but who address the causes of absence, unhappiness or underperformance among individuals…By contrast, some organisations introduce a range of benefits, but still find their people are stressed, demotivated or take frequent short-term 'sickies'."

The difference, it points out, is in: "…the concern the organisation shows for staff wellbeing across the range of people management and development activities. The behaviour of line managers and teams is key, as is the way the people are supported in balancing work and personal life."

Investors in People has developed Health and Wellbeing at Work around five challenging themes (see box on the opposite page). It has a clear emphasis on management's need to address the working experience, and to go further than legal responsibilities dictate. The draft standards look for evidence of health and wellbeing as part of business strategy and subject to performance indicators; development and training paths for staff (eg. stress management, flexible working) and equal access to wellbeing initiatives. It also assesses the capabilities of line managers in attendance management, leadership (eg. coaching, active monitoring of workloads), and in recognition and worker involvement. The new requirements being piloted will be an extension to the main Investors in People assessment.

Investors in People is developing the framework with the support of the Department of Health, and a steering group, which comprises members of other government departments, employers and stakeholders from other groups. This is a measure of the importance being attached to the project's aims,

THE INVESTORS IN PEOPLE 'CHALLENGE' TO EMPLOYERS

1. **line management and workplace culture**
   The style and capabilities of managers, their ability to manage team members effectively, address issues of attendance and rehabilitation, identify potential causes of stress, and signpost individuals to sources of help and support – as well as the team environment

2. **prevention and risk management**
   Going beyond legal requirements to manage and address risks to health, manage stress, and prevent harm to health both physical and mental

3. **individual role and empowerment**
   Ensuring the design of job roles and the nature of communication and objective setting in the organisation promotes individual wellbeing

4. **work-life balance**
   Going beyond legal requirements to support employees working flexibly where this meets the needs of the organisation and individuals

5. **enabling health improvement**
   Supporting – in a proportionate, voluntary way – employees who want to live healthily

which are to raise awareness, roll out the framework and, ultimately, to contribute to the productivity of the UK workforce.

If Investors in People looks like being a flagship for workplace health and wellbeing, there are plenty of other projects and tools that are focusing on other perspectives of a healthy organisation.

Another project with government backing is Well@Work, which aims to build the evidence base of 'what works' to promote healthy lifestyles in workplaces. It uses interventions to tackle key lifestyle behaviours, such as smoking, healthy eating and physical exercise, as well as back care, stress and mental wellbeing. These are then the subject of workplace audits and evaluation, and the results are measured against potential outcomes, including changes in health behaviours and work-related factors such as job satisfaction.

The private workplace health sector claims to already have a good idea about 'what works' to shape a healthy organisation. Vielife, a workplace health specialist, has carried out a controlled study to explore the links between health and performance. The study, conducted in conjunction with the Institute for Health and Productivity Management, found that a multi-component health promotion programme significantly improved the health status of an intervention group, which also saw an 8.5 per cent improvement in work performance. There was also less absence, and Vielife calculates a return on investment of £3.73 for each £1 spent.

The study was carried out on an intervention group from Unilever and a control group made up of the general working population. The findings reveal just what a broad range of risk factors can affect health and performance, and just how many high risk individuals there are (see table below for risk factor prevalence for the whole population). The intervention group had 17 per cent with five or more of these risk factors – and the control group 27 per cent.

In this particular study, improvements in stress, pain management and sleep were the main factors affecting health status and work performance. The study also found that the risk factors can vary greatly among departments within a company, so a breakdown by work function can provide crucial information.

HEALTH RISK FACTORS

The following table shows the baseline prevalence of the health risk factors that were assessed during the study for the population sample

risk factor	per cent population at risk
sedentary lifestyle	43
poor nutrition	34
presence of certain medical conditions	34
significant sleep problems	28
high levels of stress	24
obesity (body mass index 30 or greater)	23
significant pain	18
smoking	17
excess alcohol consumption	13
poor perception of general health	13
poor job satisfaction	12
excess sickness absence	3
seatbelt use	1

Source: Vielife

More can be found on ways of measuring workplace health in chapter 3. Numerous tools are available for assessing all or part of workplace health. Familiarity with the factors they investigate can only add to the understanding of a company's own healthy status. A good example is the HSE's Stress Management Standards (see also chapters 3 and 6).

The abundance of all these tools, standards and research studies could cause employers to lose sight of the very people – the employees – they are all aimed at. Investors in People is clear that the relationship between management and workforce is critical.

Some recent surveys have specifically targeted staff to find out what their main concerns are. Here are some of the findings:

- some 20 per cent of workers are concerned about work-related stress, and 40 per cent feel that the risk of stress can be realistically reduced, according to a recent HSE Workplace Health and Safety Survey (WHASS)

- healthy food was voted one of the most important components of a healthy lifestyle, according to a survey by disability insurer UnumProvident. Even so, 42 per cent of respondents do not have access to healthy choices. Sixty seven per cent said their employer does not help them with exercise facilities, while 43 per cent have no access to an independent person to talk to about stress or work-life issues

- a survey by web conferencing firm Webex found that two thirds of office workers are not permitted to work at home, even though one in three respondents said they would pay out of their own pockets towards flexible working technologies

- another UnumProvident survey revealed that salaries and other benefits are the least important factors in determining job satisfaction. It is the nature of work, the workplace environment and working relationships that are the most crucial for career contentment

The days when an organisation's health was defined only in terms of the bottom line are disappearing. Even today's top politicians are grappling with work-life balance. Clearly, then, managers who do not listen to their employees' health and wellbeing concerns at work choose to do so at their peril.

It could hurt his employer too

Falls caused by slips and trips cost businesses £500 million each year. They are the most common type of workplace accident. If staff were encouraged to clean up spills and keep workplaces tidy, everyone could be saved a lot of unnecessary pain.

For more information visit www.hse.gov.uk or call 0845 345 0055.

Don't just see it, sort it.

HSE

Better health & safety benefits everyone

measuring and monitoring

> **Without proper measuring and monitoring systems, a business will be unable to manage its workplace health effectively, says Stephen Bevan, director of research, The Work Foundation**

If employers are to manage the consequences of ill health in the workforce it is important to have simple systems to measure and monitor the nature and extent of the problem. This is the first step to managing it effectively and responsibly.

Unless employers have at least some idea about the patterns of illness or injury, and how these are manifested in sickness absence or under-performance, it will be difficult to target policies and practices on the areas where they will be most effective. This chapter focuses on two key workplace health management approaches – absence measurement and health profiling – and we begin by looking at absence from work.

EXECUTIVE SUMMARY

- ☐ directors should be aware of the need to measure and monitor workplace health effectively
- ☐ there are numerous methods for calculating sickness absence rates
- ☐ surveys and health profiling are commonly used to monitor workplace health
- ☐ there are basic principles that all measuring and monitoring methods should adhere to

measuring absence

There are at least 44 different ways to calculate sickness absence rates, with 14 in common use. One reason for this is that no single measure can adequately reflect the variable patterns of absence that organisations experience. In particular, most 'headline' figures mask patterns of absence that, while dominated by sporadic or short-term absences, are skewed by a growing amount of long-term absence.

So, the national average of just over seven days per employee a year does not reflect the fact that, in many organisations, 30-40 per cent of employees have no absence at all during the year. Nor does it reveal that some employees are into their third, fourth or even fifth year of long-term sickness absence.

Most employers, at the very least, calculate the number of days lost per employee each year. Another common measure is the 'lost time' rate – ie. the percentage of available days lost per year. This is calculated as follows:

$$\frac{\text{total absence (hours and or days)}}{\text{total available working time (hours or days)}} \times 100 = \text{lost time rate (\%)}$$

Let's look at a short example of how this might be applied. A group of 100 employees are contracted to work 16,100 hours during a single month. This breaks down as 100 employees x 7 hours a day x 23 working days in the month. If, among this group of workers, the total number of hours lost to absence is 575, we can calculate the lost-time rate as follows:

$$\frac{575}{16,100} \times 100 = 3.57\%$$

These measures give a view of the 'average' position, but do not reveal much about the extremes of absence. To ensure that absence figures provide the most useful insights into possible causes, at the very least these formulae should be applied to absences by:

- function or department
- location
- occupation
- job level
- gender
- age group

Whatever formula is chosen, the basic principles that employers need to adhere to when measuring and monitoring absence are:

- use measures that allow the patterns of both short-term and long-term absence to be established and understood

- use measures that allow intelligent analysis of the patterns of absence that the organisation is experiencing. The key here is to use data to identify 'hot spots', or groups of employees with higher absence 'risk'

- use measures that allow line managers to be held accountable for the absence of their staff. Many organisations suffer from under-recording of absence. Indeed, Work Foundation research suggests that only 22 per cent of employers are confident that all their absence is being recorded. This can affect the seriousness that the problem is afforded at senior levels and, as a result, the resources made available to tackle it

- collect and examine the main reasons for absence. This usually reveals a lot about any underlying workplace health issues (eg. workstation design and chronic upper limb disorders, workload and stress issues)

Effective measurement and monitoring of absence can ensure that policies and practices to prevent or reduce it can be effectively and economically targeted.

surveys and health profiling

Another approach to monitoring workplace health issues is to use one of the many approaches to surveying employees. Health and safety legislation requires employers to conduct a risk assessment for workplace health issues, including stress. There are several ways of doing this, but we will look at just two here for the purposes of illustration.

The first is the Health and Safety Executive's Stress Management Standards. In response to the rise in reported cases of work-related stress, the HSE has developed the Standards and supporting tools to enable organisations to measure their performance in managing the key causes of stress. The approach involves gathering information about current stress levels in the organisation, and working with staff to identify key causes and solutions.

As part of the Management Standards approach, HSE has developed and tested a generic stress survey. The survey helps employers to identify the most significant causes of stress in their workplace, spot groups of employees who seem to be more at risk, set priorities and target preventive action. (See also chapter 6 for more details).

There are several issues about stress that employers should note:

- it is often mis-used as a (largely unhelpful) 'umbrella' term for a range of mental health problems that may or may not be caused by work. Stress in itself is not an illness. But intense or prolonged exposure to stress can lead to mental or physical ill health

- financial or domestic difficulties can combine with work-related pressures to cause 'stress'

- people with chronic physical conditions – such as lower back pain – are more more likely to experience depression

- 'stress' varies widely between individuals – pressure at work can be a good thing, but everyone copes with it differently

An alternative approach to monitoring workplace health is to use an online health profiling tool. These can highlight a wider range of health-related issues that allow the employee to take more control over aspects of their health and lifestyle.

The Wellness Index is an online tool that employees complete confidentially that has been developed and tested by WellKom and The Work Foundation. It gathers information on the following topics:

- satisfaction with lifestyle
- coping with pressure
- attitudes towards wellness and health
- managing personal health and wellbeing issues
- attitudes towards an active lifestyle
- levels of physical activity
- mental wellbeing

- ☐ pace of life
- ☐ physical health
- ☐ 'stress'
- ☐ demographic/biometric data

On completion, each individual receives a personalised 'wellness report' that advises them on how to improve their personal wellness, and how this might improve their general wellbeing and performance at work.

The employer receives a report on the health and wellbeing of the workforce as a whole, helping them to identify health risks, to assess why some groups of staff have higher absence than others, and to identify workplace interventions that might reduce absence and improve wellness at work.

The Wellness Index is showing that employees who are more willing to take responsibility for their own health and wellbeing report less absence. It also shows that employees in poorly designed and badly-managed jobs suffer more work-related ill health.

procedures for managing absence

At its simplest, managing sickness absence is about knowing the nature and extent of the problem and then putting simple and effective mechanisms in place to manage it.

There are a number of important elements of a basic absence policy, including:

- ☐ clear procedures
- ☐ a communication strategy
- ☐ return-to-work interviews
- ☐ line manager training

Each of these will now be described briefly.

procedures

These should include the following:

1. employees should be clear that it is their responsibility to report that they are unable to come to work, to estimate the likely duration of their absence and to provide a reason for their absence

2. in cases of medium or long-term absences, line managers should maintain regular contact with the absent employee

3. informal discussion should take place between the line manager and the employee on return to work, irrespective of the duration of absence

4. a formal review should be carried out if an unacceptable pattern or level of absence continues, with possible reference to occupational health professionals or, in extreme cases, recourse to established disciplinary procedures

5. individual attendance targets should be set. Alternative working patterns or moving an employee to alternative duties can be considered

6. clear procedures and guidance should be given for self-certification of sickness absence

Many employers with such procedures have found that their very existence and consistent application can have an immediate effect on sickness levels.

communication

Any absence policy should be clearly communicated to all staff so that they are aware not only of what is required of them, but also what support may be available to them (for example, occupational health or counselling services).

Again, clarity of communication can be key to employees understanding that attendance is under scrutiny. In some organisations, absence procedures fall under the scope of formal consultation arrangements. It is often the case that trade unions are as concerned as management over unwarranted sickness absence levels, though they will also have obvious concerns over consistency in the application of procedures, especially where these lead to disciplinary action.

return-to-work interviews

Line managers should hold an interview with the employee on the day that he or she returns to work. This will emphasise the point that the period of sickness absence that has just finished – no matter how brief – has not gone unnoticed. It also provides the employee and their manager with an opportunity to discuss, informally (unless there is a recurrent problem) any ongoing or underlying problems.

line manager training

The role of line managers is crucial to developing good practice in managing attendance, since they have the closest contact with the individuals concerned. Action taken by other parties – such as the HR department – is likely to be less timely, more formal and out of touch with the detail of the circumstances.

Line managers should receive appropriate training and support in a number of areas. These include how to implement agreed procedures, how to influence factors that contribute to absence – such as working environment, aspects of morale, access to flexible working arrangements, etc. – and also how their actions can affect the health and attendance of staff.

conclusion

As employee health and wellbeing rises up the management agenda, it will be important that all employers are able to monitor and manage both the causes and consequences. While, for many firms, employee health is a risk management issue, there is also compelling evidence that it is a productivity and performance issue. Indeed, unless more firms grasp the nettle of employee health and wellbeing, it will become an issue of competitiveness too.

the business case

> Properly managed workplace health is delivering tangible financial and productivity gains to small companies, says Michael Wright, director of Greenstreet Berman, a health and safety management training and consultancy company

Why should smaller firms care about wellbeing at work? This question has been put to directors of small firms and they report that improving wellbeing brings the following business benefits:

- satisfying client requirements – winning more and better quality work
- controlling insurance premium costs
- improved productivity – through a reduction in absence rates
- being a good employer – motivating and retaining staff

EXECUTIVE SUMMARY

- wellbeing at work policies will enhance a company's reputation and can even help to win new business
- companies that have a good track record on managing potential risks enjoy better insurance premiums
- return to work interventions are proven to reduce rates of absence and lessen the risk of litigation

A key benefit of wellbeing at work is the enhancement of a company's reputation as a responsible and properly managed organisation that clients and staff want to work with, and can trust. For example, Cougar Automation, an industrial software specialist, changed its culture to make health and safety a key business driver. More specifically, it launched a stress initiative covering stress awareness, prevention and a return-to-work process. "Now, we not only have higher staff morale and lower sickness, but have seen significant benefits in the retention of existing customers," says John Purnell, regional director and health and safety officer. "It has also been a real differentiator when winning new business and helping the company to expand." The company has recently been awarded a Thames Water framework agreement for the next five years. This success, says Purnell, can be directly attributed to the company's health and safety policies, which helped to differentiate it from its competitors.

critical business success factors

The business case for wellbeing in small firms is based on sound judgement about critical business success factors. Today's customers expect ever higher standards of health, safety and responsibility. To win business and retain customers it is critical to demonstrate that you can meet your customers' standards.

Business customers do not want any poor performance by their contractors and suppliers to reflect badly on their own approach to business. Therefore, the demand for corporate business to act responsibly inevitably leads to the same demand being placed on their suppliers. Another good example is Scottish firm Data Scaffolding Services, which has invested in new scaffolding equipment and training in its erection to prevent injury and musculo-skeletal problems. In addition to demonstrating high standards to customers, companies like this Scottish firm can now point towards tangible financial and productivity benefits from properly managing workplace health. These include:

- reducing absence. Cougar Automation has seen its staff absence rate cut over a three-year period, from 11.9 days a year per employee, to 5.2

- improving productivity. Data Scaffolding Services now often uses two men instead of three to erect scaffolding, enabling the firm to remain highly competitive without cutting profit margins

- reduced insurance costs. Data Scaffolding's insurance costs have more than halved, from £36,000 a year in 2001, to £15,500 in 2005

controlling insurance costs

Since 2002, employers have been experiencing the true cost of employers' liability insurance. Before 2002, insurers' profit from stock market investments helped to offset the cost of claims. Since then, reduced investment profits have led insurers to raise premiums. At the same time, there is an ongoing trend towards higher compensation costs owing to factors such as higher costs of health care, increased legal costs and technical changes in settlement of compensation claims. This all adds up to higher premiums.

Moreover, there has been a move towards more risk-based pricing of insurance. This has the effect of polarising premiums, with higher premiums for poorly performing firms and lower premiums (or lesser increases in premiums) for better performing firms.

Dolphin Printers, a printing firm based in Poole, has kept a rein on its insurance costs. With concern over chemical exposure a particular concern within the industry, the firm has redeveloped its Control of Substance Hazardous to Health (COSHH) manual – and formed a health and safety committee with employees that successfully identified and addressed workplace health hazards.

cost-effective return to work and rehabilitation

The above trends are raising awareness of the tangible cost of work-related ill health. The response is two fold: improve health and safety to prevent illness, and improve rehabilitation and return-to-work practices to reduce the severity of cases.

There is a thin line between prevention and rehabilitation. The two main causes of work-related absence are excessive stress (leading to other conditions) and musculo-skeletal disorders. Both of these often develop slowly, with lesser symptoms appearing before the full effects become debilitating. Early detection allows cost-effective intervention to prevent a minor problem escalating – and it can help eliminate the causes. Return-to-work interventions can reduce absence and help prevent litigation claims; further, fewer claims extend into long-term disability and insurance premiums can be reduced.

For example, East Anglian Ambulance NHS Trust piloted a bio-psycho-social functional restoration programme to address significant problems with musculo-skeletal sickness and work-related accidents. The result was a cost saving of £15,000 a month and a reduction in the average number of working days lost, from 10 days per episode of absence, to six.

Rehabilitation is most cost effective when it is:

- [ ] proportionate to the severity of the case, with low-cost return-to-work interventions, such as phased return to work and workload for minor cases, and more intensive clinical interventions for more serious cases

- [ ] appropriate, with suitable forms of treatment for the condition guided by competent health professionals

- [ ] timely, with early intervention to prevent escalation of the condition and the loss of work skills/motivation among injured persons

(See also chapter 6 for more on stress and musculo-skeletal problems, and chapter 11 for long-term absence).

recruiting, motivating and retaining staff

Recruiting, motivating and retaining staff is a critical aspect of successful business, especially given today's tight labour markets and the short supply of skilled staff. This is even more true for smaller firms, where the loss of a key person can be extremely disruptive.

Caring for the wellbeing of staff has both tangible and intangible benefits. In the case of Dolphin Printers, only 15 people have left the company in the 36 years it has been operating. The company believes that this can be attributed to its caring culture. Employees feel safe and are happy working for the company, which now has 20 staff.

achieving these benefits cost effectively

There are numerous examples of organisations helping themselves. An obvious first step is to understand what you need to do to satisfy the demands of your customers and insurers, while at the same time identifying the main sources of ill health and absence in the company. From this you can prioritise the most cost-effective actions.

Wellbeing is often dependent on how we behave at work, how we manage workloads, and how quickly we respond to the signs of stress and musculo-skeletal disorders. By developing the awareness and competence of your managers and supervisors you make the best use of an existing resource. These are also the very same people who are critical to the management of health and wellbeing in the workplace.

Workers can provide real insights into what the main risks are, and how best to overcome them. They know the work and directly experience the symptoms. By involving them in the identification of problems and their resolution you can tap into a readily available pool of help that you already pay for. Moreover, the success of any wellbeing initiative is dependent on the co-operation of those you are trying to influence – the workers. By involving them from the outset your efforts are far more likely to pay off.

In many cases it is a matter of time rather than expense. There is a wealth of freely available guidance and tools designed to help small firms. Many interventions entail changes in behaviours, such as correct sitting posture. Many interventions are an integral part of good business management, such as ensuring people have the right skills to do the job – ensuring people can work productively and without the stress of under-performance. The most popular form of return-to-work intervention is a phone call from the line manager to the employee, to explore what the company can do to help them come back to work.

conclusion

The greatest investment is often in terms of the time and energy of the company's managers. The return on this investment is confidence in their ability and commitment to achieving a productive, responsible and properly managed organisation that people want to do business with and work for – key ingredients to business growth and stability.

risk assessment

> **A risk assessment is an important step in protecting your workers and your business, as well as complying with the law, advises the HSE**

A risk assessment can help you to focus on the risks that really matter in your workplace – the ones with the potential to cause real harm. In many instances, straightforward measures can readily control risks, for example ensuring spillages are cleaned up promptly so people do not slip, or cupboard drawers are kept closed to ensure people do not trip.

The law does not expect you to eliminate all risk, but you are required to protect people as far as 'reasonably practicable'. The Management of Health and Safety at Work Regulations 1999 is the cornerstone of current health and safety legislation. Regulation 3 requires all employers to undertake suitable and sufficient assessments of workplace risks, physical and mental, and implement suitable control measures to ensure the health and safety and welfare of their employees.

Remember, you also have a legal duty of care to your employees. This means you have a legal responsibility for all health and wellbeing issues at work even those that are not covered by specific laws.

The following is an edited version of the latest risk assessment guide from the Health and Safety Executive – Five steps to risk assessment. It focuses primarily on safety issues, but the risk assessment approach can also be applied to:

☐ environmental health issues, such as providing a comfortable and clean workplace

☐ mental wellbeing at work – the HSE's Stress Management Standards (see also chapter 6) provide a risk assessment approach to organisational stress placed on employees

☐ physical health – assessing the health of individuals by various tests and measures, perhaps as part of a 'wellness' programme

step 1 – identify the hazards

First, work out how people could be harmed. The following are some tips to help you identify the ones that matter:

☐ walk around your workplace and look at what could reasonably be expected to cause harm

☐ ask your employees or their representatives what they think. They may have noticed things that are not immediately obvious to you

☐ to help identify hazards, visit the HSE website, call the HSE Infoline or contact Workplace Health Connect (see also chapter 10)

☐ if you are a member of a trade association contact them. Many produce very helpful guidance

☐ check manufacturers' instructions or data sheets for chemicals and equipment as they can be very helpful in spelling out the hazards and putting them in perspective

☐ have a look back at your accident and ill-health records – these often help to identify the less obvious hazards

☐ remember to think about long-term hazards to health (eg. high levels of noise or exposure to harmful substances) as well as safety hazards

step 2 – decide who might be harmed and how

For each hazard, identify who might be harmed, as this will help you find the best way of managing the risk. There's no need to list everyone by name, but rather identify groups of people – eg. 'people working in the storeroom'. In each case, identify how they might be harmed, ie. what type of injury or ill health might occur. For example, "shelf stackers may suffer back injury from repeated lifting of boxes".

Remember, some workers have particular requirements, eg. new and young workers, new or expectant mothers and people with disabilities are among those that could be at particular risk.

Extra thought will be needed for some hazards:

☐ cleaners, visitors, contractors, maintenance workers, etc. who may not be in the workplace all the time

☐ members of the public, if they could be hurt by your activities

☐ if you share your workplace, you will need to think about how your work affects others present, as well as how their work affects your staff

step 3 – evaluate the risks and decide on precautions

Having spotted the hazards, you then have to decide what to do about them. The law requires you to do everything 'reasonably practicable' to protect people from harm. You can work this out for yourself, but the easiest way is to compare what you are doing with good practice (available on the HSE website).

So first, look at what you're already doing, think about what controls you have in place and how the work is organised. Compare this with the good practice and, finally, consider whether there's more you should be doing, by asking whether you can get rid of the hazard altogether and, if not, what should you be doing to control the risks so that harm is unlikely?

When controlling risks, apply the principles below, if possible in the following order:

☐ try a less risky option (eg. switch to using a less hazardous chemical)

☐ prevent access to the hazard (eg. by guarding)

☐ organise work to reduce exposure to the hazard (eg. put barriers between pedestrians and traffic)

☐ issue personal protective equipment (eg. clothing, footwear, goggles, etc); and provide welfare facilities (eg. first aid and washing facilities for removal of contamination)

☐ involve staff, so that you can be sure that what you propose to do will work in practice and won't introduce any new hazards

step 4 – record your findings and implement them

Putting the results of your risk assessment into practice will make a difference when looking after people and your business. Writing down the results of your risk assessment, and sharing them with your staff, encourages you to do this. If you have fewer than five employees you need not write anything down, though it is useful as a reference point if something changes at a later date.

The HSE does not expect a risk assessment to be perfect, but it must be suitable and sufficient. You need to be able to show that:

- a proper check was made
- you asked who might be affected
- you dealt with all the significant hazards, taking into account the number of people who could be involved
- the precautions are reasonable, and the remaining risk is low
- you involved your staff or their representatives in the process

Make a plan of action to deal with the most important things first. A good plan of action often includes a mixture of different things such as:

- a few low-cost or easy improvements that can be done quickly, perhaps as a temporary solution until more reliable controls are in place
- long-term solutions to those risks most likely to cause accidents or ill health
- long-term solutions to those risks with the worst potential consequences
- arrangements for training employees on the main risks that remain and how they are to be controlled
- regular checks to make sure that the control measures stay in place
- clear responsibilities – who will lead on what action, and by when

step 5 – review your risk assessment and update if necessary

Few workplaces stay the same. Sooner or later, you will bring in new equipment, substances and procedures that could lead to new hazards. It makes sense, therefore, to review what you are doing on an ongoing basis. Every year, or so, formally review where you are, to make sure you are still improving, or at least not sliding back.

Look at your risk assessment again. Have there been any changes? Are there improvements you still need to make? Have your workers spotted a problem?

Have you learnt anything from accidents or near misses? Make sure your risk assessment stays up to date.

When you are running a business it's all too easy to forget about reviewing your risk assessment – until something has gone wrong and it's too late. Why not set a review date for this risk assessment now? Write it down and note it in your diary as an annual event.

During the year, if there is a significant change, don't wait. Check your risk assessment and, where necessary, amend it. Try to think about the risk assessment when you're planning your change to give yourself more flexibility.

KEY LEGAL AND WELLBEING ISSUES

From a legal standpoint, all senior managers should be aware of the following occupational health and employee wellbeing issues:

- ☐ stress – a very important issue for any company. Stress should be 'risk assessed' in the same way as any other workplace risk. The HSE's Stress Management Standards allow employers the means to practically identify and manage the issue of workplace stress (see also chapter 6)

- ☐ ergonomics – problems arise in the workplace when employees carry out manual handling operations that result in the employee suffering back pain, repetitive strain injury and upper limb disorders. Similarly, employees who work at display screen equipment for most of their working day can also suffer ergonomic problems from incorrectly designed workstations

- ☐ passive smoking – the immediate effects of passive smoking include eye irritation, headache, cough, sore throat, dizziness and nausea. The bill to ban smoking in public places and workplaces is expected to come into effect by mid-2007 in England and Wales

- ☐ drugs and alcohol – the misuse of drugs and alcohol can lead to accidents at work, employees taking time off work and reduced productivity. As an employer you could be breaking the law if you knowingly allow drug-related activities in your workplace and you fail to act (see also chapter 8).

- ☐ bullying – commonly a factor in harassment, discrimination and abuse allegations by employees. It represents a major cause of injury to health, both physical and mental

- ☐ working hours – research by the DTI has shown British employees work some of the longest hours in Europe, with 13 per cent of employees working more than 48 hours a week. The types of sectors and occupations where long hours are a particular problem include transport and communication workers, managers and professionals

Source: law firm DWF (www.dwf.co.uk)

CHAPTER 6

managing
work-related health
problems

There are plenty of tools to help employers manage stress and musculo-skeletal disorders – the two most common types of work-related illness, says Sara Bean, freelance healthcare writer

According to Health and Safety Executive (HSE) figures for 2004-05, of the 28 million working days lost last year due to work-related ill health, more than half were attributed to musculo-skeletal disorders (MSDs) or stress.

EXECUTIVE SUMMARY

☐ stress and musculo-skeletal disorders are the greatest causes of work-related ill health

☐ a proactive approach can reduce the incidence and severity of MSDs

☐ there is a wealth of free advice and support available to employers

Other important work-related ill health conditions include lung diseases such as asthma; dermatitis and other skin diseases; diarrhoeal and other infections; and ill health related to vibration or noise. Elizabeth Gyngell, previously head of the Better Health at Work Division at the HSE, advises employers that work-related illness can be prevented by proper design of work, work equipment and health and safety procedures, sound health and safety management and efficient reporting mechanisms. These must be reinforced by ensuring that staff adhere to health and safety regulations, use appropriate protective equipment – such as masks and gloves – and follow other appropriate health and safety rules.

More advice is available on how to protect your staff against ill health caused by asthma, noise, skin conditions, infections, etc. on the HSE's website.

MSDs and stress require careful management within a workplace setting. Research shows that musculo-skeletal and stress-related problems can be caused by a combination of physical, occupational and psycho-social factors, which makes their management all the more difficult. However, a number of tools are available to help employers deal with these important health and work issues.

musculo-skeletal disorders

According to the HSE, musculo-skeletal disorders (MSDs) are the most common cause of workplace illness in the UK and account for over half of the estimated two million people who reported suffering from work-related ill health last year.

The term MSD describes three principal conditions – those that mainly affect the back, the upper limbs or neck – sometimes described as upper limb disorder or 'repetitive strain injury' (RSI) – or the lower limbs.

There is a lot that can be done to help reduce the chances of MSDs occurring in the workplace. Since MSDs cannot always be prevented, employers must encourage the early reporting of symptoms, and ensure that employees receive proper treatment and suitable rehabilitation combined, where appropriate, with an analysis and possible redesign of the work and work equipment.

prevention of MSDs

By being proactive, you can reduce risk of work-related back injuries by:

- carrying out a risk assessment, which simply means carefully examining what could cause harm to people and ensuring that you take the necessary precautions. Known risk factors include repeating a task too frequently, lifting weights, uncomfortable working positions, not taking breaks, cold environmental conditions and work-related stress

- eliminating or reducing the risks that can cause back pain by, for example, changing the way the work is organised or introducing lifting equipment

- designing the task and the workplace to take account of the risks

- reviewing the situation in conjunction with the workforce to ensure the changes are effective

During the risk assessment, ensure you consult with members of staff, including line managers and union or staff reps on the working practices and particular job functions of all the employees within the organisation.

Those employed mainly in manual handling tasks will require training in the correct way to lift and bend. A Manual Handling Assessment Chart tool can be downloaded from the HSE website. The chart provides a guide to assessing the risks posed by lifting, carrying and team manual handling activities.

An ergonomic assessment should be undertaken of computer users' workstations, including desk, chair and positioning of the computer and keyboard, and computer users should be encouraged to take regular screen breaks. It is vital also to look at the nature of the work and the workflow. The Institution of Occupational Safety and Health publishes a useful guide to ergonomics (see resources) and the HSE also provides a section on the subject.

Other practical ways of guarding against back problems include:

- ensuring loads are not handled above shoulder height or in cramped working areas

- arranging cover for holidays and unexpected absences so that no-one is left to cope alone in handling loads normally done by two or more workers

- keeping the workplace clear of obstructions that can cause trip and slip accidents when handling loads

- provide workers with information on healthy backs, such as that provided by the charity BackCare, or in the *Back Book* (see below)

managing staff with MSDs

Encouraging sufferers to stay active and continue to carry out normal activities as much as possible has been shown to be the best way to avoid long-term disability and sickness absence. When staff do take sick leave, there are ways you can help precipitate a positive return to work:

- provide a copy of the *Back Book*, published by The Stationery Office. This offers advice on how to cope with back pain and lead a normal life

☐ consult with the worker on what they might find difficult about their job and ways of making their job less physically demanding to help reassure them of a quick return to work

☐ if the condition continues longer than a few days, encourage the member of staff to see a medical practitioner – ideally an occupational physician

☐ if the member of staff does take extended sick leave, make sure you keep in touch and discuss whether modified work or a gradual build-up to normal duties will help them return to work

Back pain at work is a short guide for small businesses, produced by Working Backs Scotland, which gives advice on managing back pain and encouraging sufferers to stay active to avoid worsening the condition. Small business employers in England and Wales can also get free advice on specific workplace health concerns from Workplace Health Connect (see also chapter 10).

stress

Many managers feel out of their depth when faced with employees' mental health problems. Although this may appear a daunting prospect, when you consider a total of 12.8 million working days were lost to stress, depression and anxiety in 2004/5 and that each individual case of stress-related ill health leads to an average of 30.9 working days lost (source: HSE) it makes good business sense to take steps to combat the problem. A recent CBI/AXA survey on absence rates reveals that three quarters of firms now use stress management policies.

Stress Management Standards

The HSE has devised Stress Management Standards (see box opposite) to help employers understand the issues surrounding stress and minimise the effects on their business. The success of this approach will depend on good communication between senior managers and staff throughout the process.

Once you've familiarised yourself with the standards, you can apply them to your organisation. Use them to help identify any potential causes of stress within your company to help determine what action should be taken to address

THE STRESS MANAGEMENT STANDARDS

The six key areas identified by the standards as primary sources of stress at work are:

- [ ] **demands** – workload, work patterns and work environment
- [ ] **control** – how much say someone has in the way they do their work
- [ ] **support** – encouragement, sponsorship and resources provided by managers and colleagues
- [ ] **relationships** – whether positive working is promoted to avoid conflict
- [ ] **role** – do people understand their role within the organisation?
- [ ] **change** – how is organisational change managed and communicated?

these factors. Once you've evaluated the risk, take action by consulting with your employees to discuss problem areas and work with them to develop any further actions that may need to be taken. At this stage, ensure that any specific issues affecting particular individuals are addressed. The results of the general assessment should be fed back to employees, with a commitment to follow-up. Record your findings and agreed actions as you go along.

review and follow up

It is important to review your efforts to tackle stress by monitoring your action plan to ensure that the agreed actions – changes to working processes or agreed meetings, etc. – are taking place. It is also essential that you evaluate the effectiveness of the solutions you implement by, for example, checking absence figures and polling staff on their views.

More information on the Stress Management Standards can be found on the HSE website, while the International Stress Management Association (ISMA) has published a useful guide for employers on implementing the standards.

The HSE is currently reviewing the impact of the standards on approximately 70 companies, for publication later this year. In the meantime, if you'd like to share your experiences or find out about the practical solutions used by other organisations, log onto the HSE Stress Solutions Discussion Group (register at http://webcommunities.hse.gov.uk and navigate to the correct group).

defining diversity

Diversity is as much about age and disability as it is race and gender, says Nic Paton, freelance healthcare writer

The concept of diversity in the workplace is most readily associated with racial and gender diversity – ensuring workforces have an ethnic, gender and cultural mix that reflects their society and customer base. But diversity also encompasses disability and age.

Britain's working population is becoming older. By 2030, half the UK population will be aged over 50, and one third over 60, which means that health and disability issues will quickly move up the business agenda.

Furthermore, moves by the government to reduce the size of the welfare state by getting people who have been sick or disabled – and on benefits – back into employment is making this an issue that businesses of all sizes cannot afford to ignore.

EXECUTIVE SUMMARY

- with the working population getting older, age will become a critical issue
- perceptions need to change. It's not about obligations, but benefits
- don't assume someone is incapable of doing a job because of their age or disability. Rather, carry out a risk assessment
- solutions need not be complex, radical or costly

According to the Employers' Forum on Disability (EFoD):

- one in eight employees has a disability, the equivalent of 3.4 million people
- each year 25,000 people leave work because of an injury and ill health
- more than a third of UK businesses have hard-to-fill vacancies. Yet 3.4 million disabled people are out of work
- at least 1.5 million part-time disabled workers say they are working below their potential

the diversity 'Cinderella'

The wider message that there is a clear business benefit from having a diverse workforce appears to have got through to businesses, by and large. In June 2006 lobby group Race for Opportunity published research suggesting that the overwhelming majority of large UK employers (91 per cent) now recognise the business case for diversity.

Yet when it comes to disability the message has been much less widely assimilated. This is despite advances in legislation (most notably the 1995 Disability Discrimination Act and the creation of a Disability Rights Commission).

Last year, the EFoD last year identified disability as the diversity 'Cinderella'. It found that while 90 per cent of organisations have an allocated budget to support race equality and 68 per cent have a budget on gender equality, only 48 per cent put money aside for disability equality.

tackling perception problems

The Chartered Institute of Personnel and Development (CIPD) suggests that growing recruitment difficulties and the introduction of new anti-age discrimination laws in October 2006 has led many employers to adapt their recruitment and retention policies to target older workers.

Yet, at the same time, it warns that age discrimination persists in many organisations, with 59 per cent of workers having reported that they have been disadvantaged at work because of their age.

Straddling these two issues is the government's drive to get more workers, many of them older and disabled, off Incapacity Benefit and back into the workplace. The plan, outlined by the Department for Work and Pensions in its *Health, work and wellbeing – caring for our future* paper in 2005, and then its Green Paper *A New Deal for Welfare: empowering people to work* in January 2006, stresses the need to use occupational health, together with GP and employment services, to encourage, rehabilitate and support workers back into employment.

Yet research carried out by the CIPD suggests that employer attitudes may be one of the biggest stumbling blocks, with one in three employers saying they deliberately excluded people with such histories when recruiting staff.

opening up your business to people with disabilities

The diversity agenda can sometimes appear intimidating, particularly around disability. But it need not be, and often the solutions are not radical or expensive.

The first point to be clear on is that it usually makes more sense to retain an existing employee, even if it means adjusting their job, than it does to start again. Recruiting a new employee is estimated to cost on average around £8,200. This is before taking into account lost productivity while the new person learns the job or someone has to cover the vacancy.

The EFoD stresses that when recruiting people with disabilities, it is essential to look carefully at your processes as well as at obvious things such as physical access. People with a disability – particularly those who are visually impaired – can struggle to access online job sites or even company websites. Meanwhile, people with hearing impairments can have trouble with corporate phone systems and, too often, employers hold interviews in rooms that are inaccessible or put disabled people at a disadvantage.

Businesses need to ensure interviewers have appropriate training and are skilled in making adjustments for individual candidates. The EFoD also emphasises that when it comes to disability, employers need to aim for 'fair rights' – treating disabled people differently to ensure fairness – as opposed to equal rights, where everyone is treated absolutely equally.

The EFoD offers access to a helpline and runs a Disability Standard that allows businesses to audit their progress on disability and plan for the future. Businesses, it argues, need to develop 'disability confidence' in how they interact with both customers and potential employees.

It also has a website, www.realising-potential.org, where businesses can access support and benchmark their disability diversity performance.

Along with the Royal Association for Disability and Rehabilitation, the EFoD also operates an online access register, called Direct Enquires. For a fee of £35 a year, businesses can use this to audit their premises online and show what services they offer to people with disabilities.

taking an holistic approach to older workers

Managing and recruiting older workers need not be any more challenging than any other workers, argues the Employers Forum on Age (EFA). Older workers may have specific health needs, but can often be fitter and have more focus, enthusiasm, customer-facing skills and loyalty than many younger workers.

The key, says the EFA, is to look at the process holistically. It is not about ticking a box that shows you are good on diversity, but rather about hiring an individual who you can trust to do a job well – even if that job may need some adjustments – and who will then be able to repay the trust you have put in them.

The same principle applies to absence due to sickness of an older employee, it says. Many firms report that older workers generally take fewer sick days than their younger counterparts. While there may be issues around manual handling tasks, musculo-skeletal disorders or age-related disabilities, it is important not to make general assumptions on the basis of an individual's age. Rather, says the EFA, the company should carry out a risk assessment of the job and use it to determine the fitness of the individual employee to do it.

Having a clear, well-communicated diversity policy in place that covers both age and disability is advisable. An education programme for management could also prove useful, especially in helping younger managers not to feel threatened by the prospect of managing older workers. And, a general programme of education for all staff would ensure that there is not an 'ageist' culture within the organisation.

External sources of advice and guidance include the government's Age Positive campaign.

The EFA suggests the following steps to establishing an age-neutral work environment:

- 'age profile' your workforce across different parts of the business
- ensure that your employment, training, communications and other policies do not disadvantage particular age groups
- include age within your equal opportunity statement/policy, if appropriate
- remove unnecessary age criteria from day-to-day practice and procedure. Make age awareness training available to all employees
- ensure top-level involvement and establish an 'age champion' to keep age on the board's agenda
- help employees, customers, suppliers and the wider community to recognise your organisation's commitment on age
- create measurable performance indicators, develop accountability and establish benefits to the bottom line

let's be reasonable

The CIPD points out that those organisations that recognise that promoting diversity – in all its forms – will add value to their business (as opposed to something to be done to keep them out of court), will find the process much less onerous.

On top of this, it is much easier for a business to defend its record if it can prove it has made 'reasonable' adjustments when – or before – being asked, taken complaints seriously, worked with employees and generally been proactive in trying to bring in and support as wide a range of people as possible.

Most employees, too, if treated well are reasonable. If they are happy and confident about their job then, even if their employer is not completely perfect, they are much less likely to challenge the circumstances in which they are working. People simply want to do a job, do it well and be valued.

alcohol, drugs and smoking at work

Alcohol, smoking and drugs can have a profound effect on the workplace, says Sara Bean, freelance healthcare writer

EXECUTIVE SUMMARY

- ☐ an estimated 70 per cent of people with alcohol problems are in employment
- ☐ drug and alcohol abuse can impair performance and increase risk of accidents
- ☐ the forthcoming smoking ban in England & Wales will affect all public places
- ☐ a workplace smoking cessation programme will help both smokers and non-smokers

A survey by the government's Strategy Unit has revealed that alcohol misuse costs business up to £6.4bn a year in lost productivity through increased absenteeism, unemployment and premature death. Further evidence of substance abuse in the workplace comes from an HSE report that found that 13 per cent of working respondents (29 per cent of whom were under 30) admitted to drug use during the previous year. The findings also confirmed that there is a link between drug use and a deterioration in cognitive performance.

Meanwhile, smoking remains the largest preventable health problem in the UK, and is responsible for 34 million lost days through sickness absence annually.

Under the Health and Safety at Work etc Act 1974 (HSW Act), employers are legally responsible for addressing alcohol, smoking and drugs use at work. It is advisable, therefore, for companies to have a clear drugs and alcohol policy and to implement a smoking cessation programme. By doing so, you can be confident of helping staff and any other people who could be affected by drug and alcohol misuse, and of meeting the requirements of the forthcoming workplace-smoking ban in England, Wales and Northern Ireland.

alcohol and drugs misuse

Alcohol and substance abuse is known to:

- impair work performance

- damage customer relations

- foster resentment among employees who have to 'carry' colleagues whose work declines because of their problem

Drinking even small amounts of alcohol before or while carrying out work that is 'safety sensitive' can increase the risk of an accident. Even at blood alcohol concentrations lower than the legal drink/drive limit, alcohol reduces physical co-ordination and reaction speeds, affects thinking, judgement and mood.

Drugs often lead to impaired judgement or concentration. They can also bring about the neglect of general health and wellbeing and may adversely influence performance at work – even when the misuse occurs outside the workplace. Employers must also be aware that use – and misuse – of legally prescribed drugs can also impair reaction times and cause drowsiness. The safety implications should be considered in risk assessments.

benefits of a drug and alcohol policy

Drawing up a written drug and alcohol policy document benefits an organisation by:

- saving the cost of recruiting and training new employees to replace those whose employment might otherwise be terminated because of drugs or alcohol misuse

- reducing the cost of absenteeism or impaired productivity

- creating a more productive environment by offering support to those employees who declare a problem

- reducing the risk of accidents caused by impaired judgement

- enhancing the public perception of the business as a responsible employer

Before drawing up the policy, decide which stakeholders within the workplace

– line managers, union and staff reps, health and safety – will be consulted. Then decide what position the company wants to take on specific issues. It is also important to determine how the workforce is informed regarding a formal policy on alcohol.

drug and alcohol policy

Your policy document does not have to be long or complicated, but should comprise the following key elements:

① aims: you need to set out why the policy exists, and who it applies to. Does it apply equally to all staff or just to those in, for example, safety-critical roles?

② responsibility: the policy will be more effective if a senior employee is named as having overall responsibility

③ a definition of what constitutes drug/substance misuse

④ the rules: how does the company expect employees to behave to ensure that alcohol consumption or drug misuse do not have a detrimental effect on their work?

⑤ special circumstances: do the rules on alcohol consumption apply in all situations or are there exceptions?

⑥ safeguards: make it clear whether absence for drug or alcohol abuse treatment and rehabilitation is regarded as normal sickness. You will also need to clarify the policy on relapses and whether drug testing is applicable (see below).

⑦ confidentiality: a statement assuring employees who admit to an alcohol or drug problem that their disclosure will be treated in strict confidence

⑧ help: a description of the support available to employees with problems

⑨ information: general information about the effects of alcohol and drugs use on health and safety

⑩ disciplinary action: the circumstances in which disciplinary action will begin

drug and alcohol testing

Lindsay Hadfield, policy and education consultant at Medscreen, advises that drug and alcohol testing should only be used in conjunction with a written policy. Its introduction is likely to require a change in employee's contractual terms and should, therefore, be carefully negotiated.

There are three types of testing:

- pre-employment testing
- 'for cause' testing, which may be in response to behaviour which suggests drugs or alcohol abuse. This type of testing can be used as a deterrent in cases where a health and safety risk has already been identified
- unannounced testing. Employers who intend to use random testing need to have very clear justification for doing so – for example, where there are serious safety consequences for drug and alcohol misuse

Employers who need advice on drugs and alcohol abuse at work can speak to a workplace business trainer at the Home Office's National Workplace Initiative (see www.drugs.gov.uk). They are on hand to:

- advise and review the organisation's drug and alcohol policies
- assess and advise individual employees for treatment
- advise companies on how to promote sensible drinking

This resource can also be accessed via the internet, by going to www.drugs.gov.uk/drug-strategy/drugs-in-workplace/national-workplace-initiative

In his guide for small businesses on the Employment Practices Code to the Data Protection Act 1998, the Information Commissioner makes clear that 'collecting information by testing workers for drug or alcohol use is usually justifiable for health and safety reasons only'.

The Commissioner also advises that where testing is carried out it should meet specific good practice principles (see box on opposite page). Further guidance is available from the Information Commissioner's Office (see resources).

GUIDANCE ON DRUG TESTING AT WORK

The Information Commissioner's guidance on drug testing at work is as follows:

- only use drug or alcohol tests where they provide significantly better evidence of impairment than other less intrusive means
- use the least intrusive forms of testing that will bring the intended benefits to the business
- tell workers what drugs they are being tested for
- base any testing on reliable scientific evidence about the effect of particular substances on workers
- limit testing to those substances and the extent of exposure that will meet the purpose(s) for which the testing is conducted
- ensure random testing is genuinely random. It is unfair and deceptive to let workers believe that testing is random if, in fact, other criteria are being used
- do not collect personal information by testing all workers, whether randomly or not, if only workers carrying out a particular activity pose a risk. Workers in different jobs will pose different safety risks, so the random testing of all workers will rarely be justified

Source: The Information Commissioner's Office

rehabilitation

Notwithstanding the need for clear rules on drugs and alcohol misuse in the workplace, particularly in safety-critical situations, employers should be prepared to support employees who declare a drugs or alcohol problem. Employers with alcohol and drugs problems require specialist help and should be referred to expert support services. Guidance is available from a number of organisations including the National Treatment Agency for Substance Misuse, Alcohol Concern and Drugscope (see resources). Employers should treat information about drug or alcohol problems in strict confidence, but may need to discuss with the individual the possibility of limited disclosure to others, such as the supervisor or line manager, for example in relation to safety-critical work.

Employers will need to look carefully at the duties undertaken by the individual during the rehabilitation period, again particularly where safety is a major concern. Temporary deployment to non-safety-critical jobs may be advisable. Refer to expert help when making decisions such as these.

smoking and the workplace

Half of regular smokers will be killed by their habit. It is the primary cause of lung disease, and is associated with complications in pregnancy. Aside from causing workplace absences due to ill health, smoking also leads to lost time during the average working day. In addition, secondhand or passive smoking can damage the health of non-smokers leading to sickness, loss of productivity and the threat of litigation.

Scotland has taken the lead on banning smoking in public places, introducing the ban in March 2006. From the summer of 2007, smoking will be banned in all public places in England and Wales. This means pubs, cinemas, offices, factories and public transport will all be smoke-free. A full ban will come into force in Northern Ireland in spring 2007.

smoking cessation policy

A no-smoking policy should not be about which people smoke, but rather where, and possibly when, they smoke. It should aim to ensure non-smokers are protected from secondhand smoke and give support to those wishing to give up. Smokers who want to quit can also be directed to the national network of NHS 'Stop Smoking Services'.

CASE STUDY

Earlier this year Scottish Gas became the 1,000th company to win a national clean air award for going smoke free.

The company – part of the Centrica Group – has implemented a smoke-free environment within its buildings, which extends to entrances and exits.

As well as enforcing the ban, it has also helped many of its employees complete a smoking cessation programme and in so doing has made those staff that want to stop smoking feel supported rather than penalised. At the same time, the rules against workplace smoking have made the non-smokers feel equally considered and cared for.

The company was offered help and advice from the Roy Castle Lung Cancer Foundation, which launched the awards to help reduce public and workplace exposure to secondhand smoke. For more information, see www.roycastle.org.

It is important that employees are given time to adapt, so a three-month notice period is recommended.

When drawing up the policy, first decide whether there will be an outright smoking ban during the working day or whether provision will be made for smokers by way of designated smoking areas/shelters. Other issues to consider are visitor-smoking, support for smokers, disciplinary action for policy breaches and when the policy will be reviewed. You can download a model policy from www.cleanairaward.org.uk. Employers should start to think about how they will respond to the forthcoming smoking bans by preparing well in advance.

benefits of a smoking cessation programme

A no-smoking policy reduces legal liability, creates a safer working environment, improves workers' health, reduces tensions between smokers and non-smokers and demonstrates a caring approach to all staff and customers. It also has significant cost benefits, including:

- improved working relationships and morale

- reduced sickness and early retirements due to ill health

- reduced annual healthcare costs and health insurance for smokers

- fulfillment of health and safety regulations and reduced risk of litigation

- reduced risk of fire damage, explosions and other accidents related to smoking

- reduced maintenance and cleaning costs

- greater appeal to non-smoking customers

And, finally, a smoking cessation programme is a relatively straightforward initiative to introduce, given the support available.

flexible working

Implementing flexible working practices should be no more difficult for employers to carry out than any other workplace policy, says Howard Fidderman, freelance writer and editor of *Health and Safety Bulletin*

One of the most important employment trends in recent years has been the move towards 'flexible' forms of working that accommodate the needs of both employer and worker.

There are many types of flexibility. Essentially they embrace either different hours or a different location – or a combination of both – as compared to 'normal' working at an employer's premises. Some employers – aware of the potential benefits for productivity and improved staff mental health – have been pursuing mutual flexibility for many years. This trend received a boost in April 2003 when the government gave parents of children under six (or under 18 if disabled) the right to request changes to their working arrangements. As from April 2007, carers of adults will have the same right.

EXECUTIVE SUMMARY

☐ with work-related stress increasing, companies are introducing more flexible working options

☐ keeping communication lines open with home workers and those working flexible hours in the office is key

☐ with safeguards in place, flexible working can bring benefits to both companies and their employees

Parents can request changes to the hours, times and location of their work. Employers must give 'serious consideration' to the request but can refuse it on any of eight grounds, including cost, inability to rearrange work or recruit extra staff, detrimental effect on quality or performance of the business. Studies have found, however, that most employers have fully or partially approved the changes requested and are starting to reap the benefits.

THE BENEFITS

The benefits that companies can derive from flexible working include:

- ○ a mentally and physically healthier workforce, with reduced stress levels
- ○ reduced absence levels
- ○ a more motivated and happier workforce
- ○ enhanced recruitment and retention of staff who might not otherwise be available, such as parents with health problems or caring responsibilities
- ○ retention of organisational experience and valuable skills by offering older workers a flexible approach to work and retirement
- ○ improved productivity
- ○ a more flexible workforce
- ○ improved image or reputation

Health and safety issues have featured prominently in the flexible working debate – often as a concern, but increasingly as a beneficial outcome, particularly at a time when stress is resulting in high levels of sick leave. There can be a tendency to mystify the health and safety aspects of flexible working, particularly by consultants looking to sell services to busy directors of SMEs. In practice, however, directors should be able to manage most risks that arise from flexible working patterns just as they would any other workplace risk. The case study on page 59 demonstrates how companies can integrate this type of risk management into the organisation's overall strategy.

flexibility at the employer's premises

Generally, flexible working patterns at an employer's premises will raise the same 'physical' safety issues as more standard forms of working. The basic risk assessment should identify any particular hazards: for example, where workers 'hot-desk' or share workstations. In such instances, the equipment will need to be particularly robust because of frequent movement and adjustment.

It is important, too, to consider the security of staff who start or finish work late at night or early in the morning. Typical concerns would include whether the

DIFFERENT TYPES OF FLEXIBLE WORKING

The various types of flexible work arrangements that employers may offer staff include:

- ☐ homeworking – working from home either full or part-time
- ☐ annualised hours – a total number of hours is worked over a year, rather than a regular weekly total
- ☐ compressed hours – working normal hours over a longer day
- ☐ flexitime – a flexible start and/or finish, usually with core hours
- ☐ job-sharing – one or more people working together to cover a full-time post
- ☐ shift working – employers remain operational for longer than a normal day
- ☐ staggered hours – employees start and finish their day at different times
- ☐ term-time working – employees take unpaid leave during school holidays
- ☐ seasonal work – for example at Christmas or during a harvest
- ☐ sabbaticals – paid or unpaid break for a significant period

Source: adapted from "Flexible working: the right to request and the duty to consider", DTI

premises are close to regular public transport, whether car parking facilities and paths to and from the premises are well-lit, and if there is a need for CCTV.

The areas that are most likely to need particular attention, however, relate to arrangements and procedures. Employers should:

- ☐ hold fire drills more frequently and at different times of the day and week
- ☐ ensure they know who is on the premises at any given time
- ☐ decide whether they need additional first-aid personnel if there are times when people are working without appropriate cover
- ☐ check that flexible patterns do not prevent the worker's access to welfare facilities, such as a staff canteen, and social activities
- ☐ provide health and safety information and training for all workers
- ☐ ensure there are good systems for detecting, and encouraging workers to report, any signs of ill health

The last two issues are equally important for home workers.

MTM PRODUCTS

MTM Products is a small, Chesterfield-based manufacturer of labels, nameplates and vinyl graphics. The company appointed Ian Greenaway as managing director in 1996 to implement a major restructuring to ensure the viability of the organisation.

Greenaway took a personal and proactive approach to health and safety, with flexible working, work-life balance and staff mental health and happiness central to his approach. "The key driver is to ensure the company maintains a safe, healthy and happy workforce because it is good for business," says Greenaway. "As a small company, any absenteeism reduces productivity."

Greenaway also appointed a health and safety officer, who reports directly to him; set up an accident prevention committee; and introduced a mental health policy. The company conducts regular work-life balance and stress reviews and staff briefing sessions.

The company's 45 employees (39 full-time equivalents) currently enjoy 28 different working patterns, including flexible hours, staggered hours, condensed weeks, part-time work and home working. Greenaway also allows staff to work flexibly on an ad hoc basis, providing they have arranged cover and ensure that the job gets done. He does not allow systematic overtime working – an issue that is often overlooked in discussions of flexible working.

Potential homeworkers must complete a questionnaire, which is followed by a visit from MTM's health and safety officer. Flexibility at the workplace is addressed through conventional risk assessment and management. Overall, Greenaway says that flexible working has thrown up few health and safety issues beyond ensuring staff are not working alone and that there is adequate first-aid provision at all times. (MTM has more first aiders and appointed persons than the HSE's guidance suggests.)

Greenaway believes that MTM's approach to health and safety (especially its use of flexible working to meet the demands outside of work, without taking time off) has helped to:

- reduce absenteeism from 2.5 per cent to 1.5 per cent, to between two and three days a year – well below the national average for the private sector

- increase productivity – sales per employee have increased by 80 per cent over five years

- allowed employees an excellent work-life balance coupled with an appreciation of the business's needs

Other benefits have included: reduced corporate risk and disruption from health and safety problems (just one minor injury in the past five years); a successful company restructuring (with staff accepting multi-skilling in return for flexibility); a happy and mentally healthy workforce; improved relations with the HSE, TUC, financial stakeholders and the public; and stable employers liability insurance premiums at a time when levels are generally rising.

MTM's approach also helped it win the 2001 National Employer of the Year award for small businesses, national and regional HSE safety week awards, and resulted in its inclusion in the HSE's 2005 series of case studies of exemplary board leadership in health and safety.

working at home

Home workers face particular risks, including mental, physical and social isolation, a lack of onsite equipment and expertise, the multi-purpose uses of a home – particularly when children and animals are present – and the physical remoteness of the employer.

It is not usually feasible to visit each worker's home to carry out a risk assessment. An accepted halfway house is to ask home workers to complete a self-assessment form, and have a safety officer or line manager check the form and pursue any problems. These can include inappropriate workstations and equipment, overloaded electrical sockets and inadequate wiring and a lack of fire extinguishers and detectors.

A big limitation of self-assessment questionnaires is that they tend to concentrate on physical hazards, which lend themselves easily to a checklist. It is more difficult to elicit information on psycho-social factors and it is important, therefore, to ensure at the start that prospective home workers are mentally equipped to handle isolation. You need to determine whether they can both motivate themselves to work and be able to shut the door – literally and psychologically – on their work. Regular contact with home workers is essential, in person as well as by email or phone, and not just to talk about output.

There are enormous advantages to flexible working. With the correct safeguards, directors can use flexible working for the benefit of the organisation and the individual.

occupational health for SMEs

SMEs have resisted providing occupational health services to staff in the past, says Nic Paton, freelance healthcare writer. But studies show that they cannot afford to do so any longer

Small and medium-sized businesses have historically been much less likely than their larger counterparts to offer employees access to occupational health (OH) or other workplace health services. According to a study by insurer AXA PPP in April 2006, OH is becoming an increasingly popular benefit within companies employing more than 300 workers, but take-up rates for smaller companies are far lower. The study revealed that six out of 10 companies employing 300 to 1,000 workers provided OH. This figure rises to 71 per cent in firms with more than 1,000 staff. By contrast, only 35 per cent of firms employing 100 to 300 workers make provision, and the figure drops to 12 per cent for firms with fewer than 100 staff.

EXECUTIVE SUMMARY

- small businesses have traditionally spurned occupational health
- absence now costs employers £13bn a year
- OH interventions are proven to lower absence rates
- the NHS, HSE and private providers now all offer OH services for SMEs

It echoes findings by the manufacturers' body EEF and the Health and Safety Executive (HSE) from 2004 that concluded that, while the majority of large companies have a full-time occupational physician or one who visited regularly, a significant number of SMEs had little or no access to one.

Why is this? For the vast majority of SMEs, simply surviving takes up every working moment of the day, so an 'intangible' benefit such as workplace health inevitably drops down the list of priorities. This is compounded by a common perception that, in a cashflow tight environment, such 'add-ons' cannot be afforded.

This situation is being made worse by a shortage of OH professionals. The Royal College of Nursing and the Association of Occupational Health Nurse Practitioners between them estimate that there are only around 7,500 OH nurses working in the UK, complemented by a few hundred OH physicians, and many of these are within the NHS and public sector.

the business case for OH

Occupational health professionals, both doctors and nurses, normally have expertise in a range of areas including:

- diagnosis and treatment of work-related disease
- assessment of risks to health and advice on managing those risks
- health, including medical surveillance
- fitness-for-work issues
- advice on pre-employment, sickness absence and ill-health retirement
- providing health education, advice on rehabilitation after illness and injury and other counselling
- first aid

Absence or illness is a growing cost, affecting all businesses, large and small. The latest survey by the Confederation of British Industry has estimated that absence costs the UK economy more than £13bn a year or around £531 per employee.

While, at 4.2 days per employee, smaller businesses generally report lower levels of absence (mostly because there is more direct contact between employees and senior staff), it is still an ongoing drain on the business.

There is also the potential legal cost of failing to manage the good health of your workers. Should an employee be injured or become ill as a result of their work, it can prove very costly at tribunal. Being able to show you have been proactive in bringing in OH expertise can go a long way towards mitigating this risk.

A survey carried out by EEF in May 2006 reported a direct correlation between providing OH and reduced rates of absence among workers, in all sizes of business. Where there was OH support (whether internally or externally provided), 39 per cent of the companies polled saw a reduction in short-term absence, while 28 per cent reported a reduction in long-term absence.

how to access OH

Small businesses can access third-party occupational health information, advice and services through:

- [ ] their local GP

- [ ] the NHS

- [ ] the HSE

- [ ] health and safety bodies

- [ ] private insurers and health providers

Taking each option in turn, the first port of call for a small business should be its local GP. It is becoming more common for GPs to have at least some understanding of workplace health issues. If your GP has expertise in this area, it may be possible to agree a contract where the doctor comes into the workplace or provides advice remotely on a regular basis.

Then there is the NHS, through its occupational health brand NHS Plus. This is a network of NHS occupational health departments across England supplying services to public and private NHS employers and, in particular, SMEs. There are currently around 90 NHS Plus providers. They can provide advice and on-site services to businesses, with services normally charged at a cost per employee, with additional travel costs.

In February 2006, the HSE launched a free confidential service called Workplace Health Connect. This is designed to give tailored, practical advice on workplace health, safety and return-to-work issues to smaller businesses (up to a

MALTHOUSE ENGINEERING

Malthouse Engineering, a West Midlands flame cutting and grinding firm, has used an occupational health provider for a number of years to deal with the health needs of its 80 employees. "It provides us with first aid, audiometry and lung function tests. We can get the nurse in daily or once a week," explains health and safety manager John Jackson.

Earlier this year, the company also approached Workplace Health Connect to see how its service could add to what it was already providing its employees. The service offered a phone consultation, site visit, discussions with employees, a follow-up report, access to an information line and educational DVDs on a range of health and safety issues.

The company has since changed its induction processes to include training on risk assessment and best use of visual display units for new recruits. "It has definitely raised morale on the shopfloor and the information line has been fantastic. It is free and the chaps feel we are trying to give them a safer and better working environment," says Jackson.

maximum of 250 workers). Any small business in England and Wales can access the advice line (0845 609 6006) and website (www.workplacehealth connect.co.uk).

In addition, businesses located in five pilot areas can arrange for a free problem-solving workplace visit by one of their expert advisers. Visits are currently available in Greater London, the North East of England, the North West, South Wales and West Midlands.

While the main attractions are that the service is confidential, free with no catches, and easy to access (see case study above), availability of the workplace visits is still limited. The service only has funding until February 2008, after which a decision will be made on whether to roll the service out nationally.

Other potential options include Safety Groups UK, the Royal Society for the Prevention of Accidents' network of 70 occupational health and safety groups around the UK. These can offer a wealth of experience and advice on health and safety at work. The Institution of Occupational Safety and Health also publishes a risk management toolkit for SMEs.

Finally, some of the big-name commercial OH providers are getting in on the act. Bupa Wellness, for instance, launched an SME service in 2005 called Health at Work. It offers a telephone and email-based helpline and services, such as

ACTION POINTS

- [ ] decide what you want outside support to do and what sort of external provider you need
- [ ] work out what results you want to achieve, your budget, target for return on investment, and what data you will be working from
- [ ] revisit this at regular intervals to monitor how successful the intervention has been
- [ ] when choosing a specialist, look at formal qualifications and areas of expertise to ensure they mesh with your needs
- [ ] talk to other clients of the provider and seek references before agreeing a contract

issuing guidance, drawing up employee health reports and screening and rehabilitating employees. It operates on an annual subscription basis, based on the number of employees.

how to decide what is right for you

Developing internal expertise – perhaps through your HR department – has obvious cost and ease-of-access benefits. However, if you do not have a manager who is competent or has the time to tackle health and safety and health and wellbeing, going the third-party route is the obvious option.

The HSE advises documenting what you want outside support to do, and what your budget and target return on investment are. Are you, for instance, looking for a one-off solution to a specific problem, or seeking a longer-term, ongoing relationship? It is also important to make sure any data you are using on absence or ill health is as robust and up-to-date as possible. It is a good idea to revisit this plan at regular intervals once the contract is up and running.

When choosing a specialist it is advisable to look at their formal qualifications and areas of expertise. It is worth asking, too, whether you actually need an OH practitioner. Ergonomists or physiotherapists may be an option (although for a more strategic, holistic approach occupational health is probably the best bet).

Another option, for companies that have regular contracts with a larger business, is to investigate the possibility of your 'piggy backing' their OH provision, suggests the Royal Society for the Prevention of Accidents.

long-term absence and return to work

Prolonged absences and return to work both need to be closely managed, and require regular communication between employer and employee, says Dr John Ballard, editor of *Occupational Health at Work* and director of The At Work Partnership

Although there is no formal definition of 'long-term absence', the term is generally recognised as referring to those that last more than 20 working days.

In terms of the way they are managed, long-term absences also include those caused by conditions that are chronic, recurrent or require repeat treatment (such as asthma or heart conditions) but don't necessarily extend beyond the four-week 'trigger point'. The most common causes are back pain, other musculo-skeletal conditions, common mental health disorders – such as anxiety, depression and ill health brought on by excessive stress – as well as recovery from accidents or surgery.

Regardless of the cause of the absence, the longer someone is off work, the lower the chances they have of returning. Anyone off work for between four and 12 weeks has a 10-40 per cent chance of still being off work after one year. Once they have been off sick for more than a year their chances of ever returning are slim.

EXECUTIVE SUMMARY

- [ ] long-term absence accounts for just six per cent of absence spells, but makes up 33 per cent of private sector and 51 per cent of public sector total lost time

- [ ] the longer someone remains on sick leave, the less likely they are to return to work

- [ ] it is important to tackle all the psycho-social, workplace and organisational obstacles that can delay or prevent return to work

- [ ] employers must be prepared to make adjustments to the work and workplace

Since most long-term ill health conditions are more common in older people, their prevalence at work will increase as the average age of the workforce rises. Employers need to think carefully about how they will manage this trend.

returning to work

How can employers work with their employees to prevent the slide from ill health to prolonged absence and incapacity? What should an employer do when an employee goes off sick with a long-term health condition? When should they intervene? And, what resources can they use to promote rehabilitation and return to work?

These questions are relatively straightforward, but successful intervention requires an integrated understanding of health and recovery and a move away from approaches that rely on medical interventions alone.

Traditional approaches to long-term absence assume that what determines whether an employee comes back to work or not – and how quickly they return – is dependent on them making a full medical recovery. But this 'medical model' does not take account of the fact that many people with chronic conditions do not recover fully. It assumes that the only intervention open to employers would be to provide quicker access to healthcare (such as a private medical plan). The medical-only approach is the reason why many people who go on sick leave never return to work, remain on incapacity benefits or take early medical retirement.

The modern approach to rehabilitation embraces appropriate medical intervention, but takes account of three non-medical 'obstacles' to recovery:

psycho-social obstacles

These include:

- the person's negative beliefs about their condition and the notion that it could be made worse by their work
- the idea that they must be free of pain or other symptoms before even thinking about work

☐ occasionally, unhelpful advice from GPs, including unnecessarily extended sickness certification

workplace obstacles

These include:

☐ the worker no longer being able to carry out all of their original duties or being unable to access the worksite, but otherwise being fit to work

☐ a lack of appropriate adjustments to work equipment, tasks or the work environment to take account of the employee's health condition or disability

organisational obstacles

These include:

☐ employers insisting that the individual is 100 per cent fit before returning, or assuming that they will be a burden in terms of absence and productivity or a risk to safety

☐ lack of appropriate contact with the absent employee

☐ poor case management, such as doing nothing until a person's occupational sick pay or permanent health insurance has expired

☐ lack of resolution in employer-employee disputes

By breaking down these obstacles – which are the same for all common long-term health conditions – and taking early action, employers can support their employees through ill health and ensure their timely return and job retention.

action plan

This inclusive approach to long-term absence management requires interventions to address all the possible obstacles to return to work as well as appropriate healthcare. But employers need to work with the absent employee to support their rehabilitation. The following is a non-exhaustive list of actions that employers can take to promote effective return to work.

☐ implement effective absence reporting and monitoring systems so that interventions are not unnecessarily delayed or misplaced

☐ train managers to understand the consequences of ill health and what they can do to help overcome the obstacles to recovery. Ensure that they don't demand that the employee is 100 per cent fit before they return to work

☐ make early and positive contact with employees absent from work. The initial emphasis should be on supporting the absent employee, enquiring about their health and wellbeing and whether anything can be done to help, rather than on return-to-work dates or pressurising them to return. Remember that not everyone recovers from illness or injury at the same rate, so be flexible. The motivation for staying in touch with employees off sick is not – or should not be – harassment, but an essential part of the employer-employee relationship

☐ where possible, encourage the employee to come in to the workplace before formally returning, even if it is just for lunch or a cup of tea. This helps the employee to remain part of the team, and prevents them becoming isolated. Maintain contact during the absence, make the employee feel valued and keep them up to date with news and developments

☐ in certain circumstances, consider funding appropriate healthcare. This might mean brief interventions such as physiotherapy for many musculo-skeletal conditions or cognitive behavioural therapy for common mental health problems. Both can be cost-effective treatments but are not always readily available on the NHS

☐ examine the workplace, equipment, tasks and working hours and consider adjustments that will help the employee return to work. Remember, employees may not be able to perform all of their previous tasks. They may require temporary changes to their work and hours so that they can return before their recovery is complete, or permanent changes if they have developed a long-term condition or disability. Individuals may require help in getting to and from work, again on a temporary or permanent basis

☐ return-to-work interviews are important in reintegrating the employee, but can also be essential in finding out if there are any adjustments required to ensure that the employee remains supported in work and not at increased

LEGAL ISSUES

- [ ] the Disability Discrimination Act 1995 (DDA) covers mental as well as physical health. It gives protection to anyone who has a health condition that is likely to last more than a year and which has a substantial and adverse impact on their ability to carry out normal day-to-day activities, or who has cancer, MS or HIV

- [ ] the DDA protects those with a disability from less favourable treatment, and requires employers to make reasonable adjustments to employment practices and premises

- [ ] employers should avoid getting into legal arguments about whether someone is or is not covered by the law and manage disability as a matter of good business practice

- [ ] when managing employees on long-term sick leave, employers must also comply with other legal duties regarding unfair dismissal, medical records and confidentiality

risk of going off sick again. This is crucial in cases where the work itself might have contributed to the absence – for example, in stress-related illness. Treat seriously any suggestions of bullying or harassment either before a person went off sick or after they have returned

- [ ] don't let an employee's workload pile up in their absence

- [ ] address genuine safety concerns if employees return to work with a health condition or on medication, but don't use health and safety as a false excuse not to accommodate them

- [ ] don't make assumptions about the individual's ill-health condition or draw unsubstantiated conclusions that they are malingering. Seek proper occupational health advice from qualified specialists. Expert 'case managers' can be called in to coordinate the return-to-work programme where the recovery process is complicated or where input is needed from a variety of resources. Occupational health professionals and case managers can work with the individual to address any negative beliefs they may have about their condition or return to work

- [ ] respect medical confidentiality. It is not a manager's responsibility to probe an individual's medical history

Return to work is not just the object of rehabilitation; it is an essential part of the recovery process. Employers who take an integrated approach to long-term absence management will reap the benefits.

looking ahead

SMEs can look forward to 'lighter' regulation and less red tape, while taking an increasingly holistic approach to health and wellbeing in the workplace, says Howard Fidderman

What trends and regulations can UK businesses, especially smaller firms, look forward to over the next few years? The good news is that the period until 2010 should be increasingly suited to the health and safety needs of SMEs. Politically, there is a move to regulating only when it is necessary and to taking a targeted, risk-based approach to enforcement.

At the same time, the government is turning to the health, work and wellbeing agenda to help deliver improvements in overall sickness absence; it also expects board-level direction of these issues to be the norm and integrated into business and corporate social responsibility agendas.

EXECUTIVE SUMMARY

- ☐ government is looking to firms to take a more holistic approach to workplace health
- ☐ government wants to reduce the administrative burdens on business
- ☐ directors will increasingly have to show leadership on health and safety issues

These developments are, in part, a pragmatic response to the need to remain competitive in a global 24/7 economy (see box) – an economy that will comprise fewer 'jobs for life' and reduced security of employment, with increased use of outsourcing and temporary contracts amid a tighter labour market. These are factors that, wrongly handled, can lead to low levels of employee morale, wellbeing and mental health. In turn, these can result in the loss of key staff, increased absence and recruitment costs.

HEADLINE ISSUES

- occupational health and safety strategies will have to address increasing globalisation and competitiveness, 24-hour working, enhanced communications technology, outsourcing, greater numbers of SMEs and a continued move from manufacturing to services

- labour force changes with workplace health implications include an ageing workforce, flexible working arrangements, increasing use of temporary and self-employed workers and workers who do not speak English as a first language

- there will be a shift in emphasis to regulating only when necessary and to taking a targeted risk-based approach to enforcement

- employers will be expected to take an increasingly holistic approach to occupational health and wellbeing

- directors will be expected to 'lead' on workplace health and wellbeing

- SMEs will have greater access to free occupational health and safety help

- there will be important new legislation, but the amount will decline

As noted above and elsewhere in this guide, the government also expects employers to see work in a wider context and to act in ways that benefit their workers, society and the wider economy, as well as the business itself. This increasingly holistic approach encompasses issues as diverse as health education, absence management, competitiveness, flexible working, quality, the environment and social responsibility.

Evidence of this can be found in specific developments such as the 2007 smoking ban (see chapter 8) and, more widely, in the government's paper, *Health, work and wellbeing – caring for our future*, which aims to pull together different government initiatives around the issues. Linked to this, the Welfare Reform Bill sits at the core of the government's ambitious plans to return one million incapacity benefit recipients back to work over 10 years. Implemented properly, a holistic approach will enable employers to attract and retain a happier, healthier and more skilled and experienced workforce.

This agenda is not an add-on; it is a core business activity that will survive any change in government, either within the Labour Party or between political parties after the next election.

INFLUENCING DIRECTORS

Directors of SMEs can now expect far better (and free) help with workplace health. Two important initiatives launched in early 2006 are the Better Business campaign – an HSE project to raise awareness among SMEs of the financial and personal costs and causes of workplace incidents, and Workplace Health Connect. (Also see chapter 10.)

---

the Hampton effect

The government has begun the task of implementing the recommendations of the Hampton report (Reducing administrative burdens: effective inspection and enforcement (final report), HM Treasury), including taking the Legislative and Regulatory Reform Bill and the Deregulation Bill through Parliament. Together, they will allow the repeal of 'unnecessary' legislation, a reduction in administrative 'burdens' on employers, a 'lighter' regulatory touch and a revised regime of enforcement penalties. The HSC/E have already disseminated their thoughts on 'sensible health and safety', and will publish a regulatory 'simplification plan' and new 'sensible risk principles' before the end of 2006.

Although Hampton is one reason why there will be less new occupational health and safety legislation, the European Commission has also accepted that it should be exploiting the Directives already in place, rather than introducing wholesale additions. Nevertheless, directors will still face important new initiatives (see box on opposite page).

corporate manslaughter

The topic that directors are most likely to be aware of is the proposed new offence of corporate manslaughter, which will apply to fatal health conditions as well as to fatal injuries.

The Bill, which was finally published in July 2006, should partly redress the unfair restriction of successful manslaughter convictions to SMEs, where it is easier to identify and convict (and jail) a director or other 'controlling mind'. The government has steadfastly ruled out sanctions against directors under the new offence (although campaigners and unions are still pressing the government to change its mind during the Bill's parliamentary progress). They

---

KEY UPCOMING LEGISLATION & STANDARDS

☐ an ongoing programme will see the roll out of the HSE's Stress Management Standards to organisations with 250 or more employees

☐ The Hampton-inspired Macrory Review allows for a wider range of penalties, some of which could be directed at directors and they range from fixed and variable administrative penalties imposed by a regulator (for example, the HSE or a local authority) to restorative justice. Criminal prosecutions in a court would, as now, be restricted to serious breaches where there have been either intentional reckless gross negligence or serious consequences

---

will, however, remain liable under existing individual manslaughter offences and, as before, directors of SMEs rather than larger organisations will be most prone to conviction.

Manslaughter may be a headline story for directors, but other initiatives will have a more substantial impact. The revised Turnbull internal control guidance on the Combined Code, published in October 2005, highlighted the importance of risk management. The HSE is currently drafting new occupational health and safety guidance for directors. Furthermore, recognising the links between health and safety and wider business needs, it is currently working with business and the IoD on major new guidance for directors on how to lead on health and safety.

the changing workforce

Looking more broadly, businesses cannot escape the fact that the UK has an ageing population and workers will have to work until later in life to support their retirement. To remain competitive, employers must adapt to this change, value the skills that older workers have and strive to keep all their employees safe, healthy and well. This should be regarded as a key business issue, with its potential to keep workers at work, benefiting themselves, their employing organisations and UK prosperity as a whole.

The following is a compilation of the resources and contacts noted in other chapters in this guide, plus other useful sources of workplace health and wellbeing information and advice.

government

Cross-government Health, work and wellbeing strategy
www.health-and-work.gov.uk

Health and Safety Executive (HSE)

The main agency responsible for providing guidance and research on workplace health and safety is the HSE. There are many more resources on its website than we can list, but key ones are:

Business Benefits of Health and Safety (includes case studies)
www.hse.gov.uk/businessbenefits

Directors' Responsibility for Health and Safety
www.hse.gov.uk/pubns/indg343.pdf

Five steps to risk assessment
www.hse.gov.uk/pubns/raindex.htm

Health and safety in small businesses
www.hse.gov.uk/smallbusinesses

Managing sickness absence and return to work
www.hse.gov.uk/sicknessabsence

Musculo-skeletal disorders and the Manual Handling Assessment Chart (MAC) Tool
www.hse.gov.uk/msd

Stress Management Standards and stress topic pages
www.hse.gov.uk/stress

worker involvement
www.hse.gov.uk/involvement

Working time regulations
www.hse.gov.uk/contact/faqs/workingtimedirective.htm

Workplace Health Connect (for SMEs)
www.workplacehealthconnect.co.uk

Department of Health

Choosing Health, the public health White Paper, can be found at:
www.dh.gov.uk

Healthy Workplace Initiative – jointly sponsored by the Department of Health and the HSE, this site provides news about workplace health, events, links to resources and a downloadable newsletter
www.signupweb.net

NHS Plus – network of NHS occupational health departments across England, supplying services to non-NHS employers.
www.nhsplus.nhs.uk

Department for Work and Pensions (DWP)

Age Positive – a campaign to end age discrimination in the workplace.
www.agepositive.gov.uk

Information on the Disability Discrimination Act
www.dwp.gov.uk/employers/dda

Department of Trade and Industry (DTI)

Business link – among the resources in the Health, safety, premises section is a Health and safety performance indicator tool (developed by the HSE)
www.businesslink.gov.uk

Flexible working: the right to request and the duty to consider: a guide for employers and employees
Available at www.dti.gov.uk

Home Office

Drugs in the workplace
www.drugs.gov.uk/drug-strategy/drugs-in-workplace

key bodies

Association of British Insurers
www.abi.org.uk

Association of Insurance and Risk Managers
www.airmic.com

British Occupational Health Research Foundation
www.bohrf.org.uk

Business in the Community (Action on Health is among its programmes)
www.bitc.org.uk

Centre for Workplace Health – established in 2005 at the Health and Safety Laboratory, University of Sheffield
www.hsl.gov.uk/cwh

Chartered Institute of Personnel and Development (CIPD)
www.cipd.co.uk

Confederation of British Industry (CBI) – publisher of the annual absence survey
www.cbi-org.uk

Disability Rights Commission
www.drc.org.uk

Engineering Employers Federation (EEF) – manufacturers' body with many health and safety resources, including a stress management guide and stress assessment tool
www.eef.org.uk

European Agency for Safety and Health at Work
http://agency.osha.eu.int

Faculty of Occupational Medicine, Royal College of Physicians – publications include Guidance on Alcohol and Drug Misuse in the Workplace
www.facoccmed.ac.uk

Faculty of Public Health, Royal College of Physicians
www.fph.org.uk

Future Work Forum
www.henleymc.ac.uk/fwf

Health Education Board for Scotland
www.hebs.scot.nhs.uk

Institute of Risk Management
www.theirm.org

Institution of Occupational Safety and Health (IOSH)
www.iosh.co.uk

Investors in People – 'Keeping people well' section in portal site
www.investorsinpeopledirect.co.uk

Scottish Centre for Healthy Working Lives (see also Safe and Healthy Working, aimed at SMEs – www.safeandhealthyworking.com)
www.healthscotland.com/hwl

Society of Occupational Medicine
www.som.org.uk

The Work Foundation – think tank on work issues with strong health and wellbeing practice
www.theworkfoundation.com

TUC (Trades Union Congress) – large repository of information in all workplace health and safety topics, and produces weekly health and safety updates.
www.tuc.org.uk

private providers

These companies (sample only) provide services such as health surveys, employee assistance programmes (eg counselling services), wellness programmes, occupational health and workplace health research. See also the Employee Assistance Programme Association (www.eapa.org.uk)

At Work Partnership – publications and training in occupational health
www.atworkpartnership.co.uk

Atos Origin
www.atosorigin.co.uk

Greenstreet Berman – risk management consultancy and developer of health and safety performance indicators for the HSE
www.greenstreet.co.uk

Grendonstar – company alcohol and drug testing
www.grendonstar.co.uk

ICAS
www.icasworld.com

Medscreen
www.medscreen.com

Standard Life Healthcare
www.standardlifehealthcare.co.uk

UnumProvident Centre for Psychosocial and Disability Research
www.cf.ac.uk/psych/cpdr

Vielife
www.vielife.com

Wellkom – is part of an initiative to construct a wellness management portal site
at www.wellnessmanagementcommunity.net
www.wellkom.co.uk

Workforce Logistics
www.workforce-logistics.com

other bodies/sites by theme

diversity

Employers for Childcare
www.employersforchildcare.org

Employers' Forum on Age
www.efa.org.uk

Employers' Forum on Disability
www.employers-forum.co.uk

Equal Opportunities Commission
www.eoc.org.uk

Opportunity Now – campaign for women workers that works with employers
www.opportunitynow.org.uk

Scope
www.scope.org.uk

flexible working/work-life balance

Employers for Work-life Balance – site run by the Work Foundation
www.employersforwork-lifebalance.org.uk

Flexibility – resource site for flexible working
www.flexibility.co.uk

Parents at Work – information, rights and research for working parents
www.parentsatwork.org.uk